Contractor's Job Book

Keep track of client information,
hours worked, and material costs

Writing Journal

Published by:
Berhampore Press
Wellington, NZ
Copyright 2017
All Rights Reserved

BerhamporePress@gmail.com

ISBN-13:
978-1544287874

ISBN-10:
1544287879

CLIENT NAME:				
ADDRESS:				
HOME PHONE: CELL PHONE:		EMAIL: OTHER:		

DATE	MATERIALS	TIME	$ HOUR	TOTAL $

COMPLETION DATE:	SUB TOTAL:
INVOICE NUMBER:	TAX:
DATE SENT:	
DATE PAID:	INVOICE TOTAL:

Notes:_____
_____
_____
_____
_____
_____
_____
_____
_____
_____
_____
_____
_____
_____
_____
_____
_____
_____

CLIENT NAME:

ADDRESS:

HOME PHONE: EMAIL:
CELL PHONE: OTHER:

DATE	MATERIALS	TIME	$ HOUR	TOTAL $

COMPLETION DATE: SUB TOTAL:

INVOICE NUMBER: TAX:

DATE SENT:

DATE PAID: INVOICE TOTAL:

Notes:

CLIENT NAME:				
ADDRESS:				
HOME PHONE: CELL PHONE:		EMAIL: OTHER:		
DATE	MATERIALS	TIME	$ HOUR	TOTAL $

COMPLETION DATE:	SUB TOTAL:
INVOICE NUMBER:	TAX:
DATE SENT:	
DATE PAID:	INVOICE TOTAL:

Notes:_____
_____
_____
_____
_____
_____
_____
_____
_____
_____
_____
_____
_____
_____
_____
_____
_____
_____
_____
_____
_____
_____

CLIENT NAME:				
ADDRESS:				
HOME PHONE: CELL PHONE:		EMAIL: OTHER:		

DATE	MATERIALS	TIME	$ HOUR	TOTAL $

COMPLETION DATE:	SUB TOTAL:
INVOICE NUMBER:	TAX:
DATE SENT:	
DATE PAID:	INVOICE TOTAL:

Notes:_____

_____

_____

_____

_____

_____

_____

_____

_____

_____

_____

_____

_____

_____

_____

_____

_____

_____

_____

_____

CLIENT NAME:				

ADDRESS:

HOME PHONE: **EMAIL:**
CELL PHONE: **OTHER:**

DATE	MATERIALS	TIME	$ HOUR	TOTAL $

COMPLETION DATE: **SUB TOTAL:**

INVOICE NUMBER: **TAX:**

DATE SENT:

DATE PAID: **INVOICE TOTAL:**

Notes:_____
_____
_____
_____
_____
_____
_____
_____
_____
_____
_____
_____
_____
_____
_____
_____
_____
_____
_____
_____
_____

CLIENT NAME:				
ADDRESS:				
HOME PHONE: CELL PHONE:		EMAIL: OTHER:		

DATE	MATERIALS	TIME	$ HOUR	TOTAL $

COMPLETION DATE:	SUB TOTAL:
INVOICE NUMBER:	TAX:
DATE SENT:	
DATE PAID:	INVOICE TOTAL:

Notes:_____

CLIENT NAME:				

ADDRESS:				

HOME PHONE: CELL PHONE:		EMAIL: OTHER:		

DATE	MATERIALS	TIME	$ HOUR	TOTAL $

COMPLETION DATE:	SUB TOTAL:
INVOICE NUMBER:	TAX:
DATE SENT:	
DATE PAID:	INVOICE TOTAL:

Notes:

CLIENT NAME:				

ADDRESS:				

HOME PHONE: CELL PHONE:		EMAIL: OTHER:		

DATE	MATERIALS	TIME	$ HOUR	TOTAL $

COMPLETION DATE:	SUB TOTAL:
INVOICE NUMBER:	TAX:
DATE SENT:	
DATE PAID:	INVOICE TOTAL:

Notes:_____

CLIENT NAME:				
ADDRESS:				
HOME PHONE: CELL PHONE:		EMAIL: OTHER:		

DATE	MATERIALS	TIME	$ HOUR	TOTAL $

COMPLETION DATE:	SUB TOTAL:
INVOICE NUMBER:	TAX:
DATE SENT:	
DATE PAID:	INVOICE TOTAL:

Notes:

CLIENT NAME:

ADDRESS:

HOME PHONE: **EMAIL:**
  CELL PHONE: **OTHER:**

DATE	MATERIALS	TIME	$ HOUR	TOTAL $

COMPLETION DATE: **SUB TOTAL:**

INVOICE NUMBER: **TAX:**

DATE SENT:

DATE PAID: **INVOICE TOTAL:**

Notes:

CLIENT NAME:

ADDRESS:

HOME PHONE: EMAIL:
 CELL PHONE: OTHER:

DATE	MATERIALS	TIME	$ HOUR	TOTAL $

COMPLETION DATE: SUB TOTAL:

INVOICE NUMBER:
                              TAX:
DATE SENT:

DATE PAID: INVOICE TOTAL:

Notes:

CLIENT NAME:				

ADDRESS:				

HOME PHONE: CELL PHONE:		EMAIL: OTHER:		

DATE	MATERIALS	TIME	$ HOUR	TOTAL $

COMPLETION DATE:	SUB TOTAL:
INVOICE NUMBER:	TAX:
DATE SENT:	
DATE PAID:	INVOICE TOTAL:

Notes:

CLIENT NAME:				
ADDRESS:				
HOME PHONE: CELL PHONE:		EMAIL: OTHER:		

DATE	MATERIALS	TIME	$ HOUR	TOTAL $

COMPLETION DATE:	SUB TOTAL:
INVOICE NUMBER:	TAX:
DATE SENT:	
DATE PAID:	INVOICE TOTAL:

Notes:

CLIENT NAME:				
ADDRESS:				
HOME PHONE: CELL PHONE:		EMAIL: OTHER:		

DATE	MATERIALS	TIME	$ HOUR	TOTAL $

COMPLETION DATE:	SUB TOTAL:
INVOICE NUMBER:	TAX:
DATE SENT:	
DATE PAID:	INVOICE TOTAL:

Notes:_____

_____

_____

_____

_____

_____

_____

_____

_____

_____

_____

_____

_____

_____

_____

_____

_____

_____

_____

_____

_____

_____

CLIENT NAME:				
ADDRESS:				
HOME PHONE: CELL PHONE:		EMAIL: OTHER:		
DATE	MATERIALS	TIME	$ HOUR	TOTAL $

COMPLETION DATE:	SUB TOTAL:
INVOICE NUMBER:	TAX:
DATE SENT:	
DATE PAID:	INVOICE TOTAL:

Notes:

CLIENT NAME:				
ADDRESS:				
HOME PHONE: CELL PHONE:		EMAIL: OTHER:		

DATE	MATERIALS	TIME	$ HOUR	TOTAL $

COMPLETION DATE:	SUB TOTAL:
INVOICE NUMBER:	TAX:
DATE SENT:	
DATE PAID:	INVOICE TOTAL:

Notes:

CLIENT NAME:

ADDRESS:

HOME PHONE: EMAIL:
 CELL PHONE: OTHER:

DATE	MATERIALS	TIME	$ HOUR	TOTAL $

COMPLETION DATE: SUB TOTAL:

INVOICE NUMBER:
                              TAX:
DATE SENT:

DATE PAID: INVOICE TOTAL:

Notes:

CLIENT NAME:				
ADDRESS:				
HOME PHONE: CELL PHONE:		EMAIL: OTHER:		

DATE	MATERIALS	TIME	$ HOUR	TOTAL $

COMPLETION DATE:	SUB TOTAL:
INVOICE NUMBER:	TAX:
DATE SENT:	
DATE PAID:	INVOICE TOTAL:

Notes:

CLIENT NAME:				
ADDRESS:				
HOME PHONE: CELL PHONE:		EMAIL: OTHER:		

DATE	MATERIALS	TIME	$ HOUR	TOTAL $

COMPLETION DATE:	SUB TOTAL:
INVOICE NUMBER:	TAX:
DATE SENT:	
DATE PAID:	INVOICE TOTAL:

Notes:

CLIENT NAME:				
ADDRESS:				
HOME PHONE: CELL PHONE:		EMAIL: OTHER:		

DATE	MATERIALS	TIME	$ HOUR	TOTAL $

COMPLETION DATE:	SUB TOTAL:
INVOICE NUMBER:	TAX:
DATE SENT:	
DATE PAID:	INVOICE TOTAL:

Notes:

CLIENT NAME:				
ADDRESS:				
HOME PHONE: CELL PHONE:		EMAIL: OTHER:		

DATE	MATERIALS	TIME	$ HOUR	TOTAL $

COMPLETION DATE:	SUB TOTAL:
INVOICE NUMBER:	TAX:
DATE SENT:	
DATE PAID:	INVOICE TOTAL:

Notes:

CLIENT NAME:				
ADDRESS:				
HOME PHONE: CELL PHONE:		EMAIL: OTHER:		

DATE	MATERIALS	TIME	$ HOUR	TOTAL $

COMPLETION DATE: INVOICE NUMBER: DATE SENT: DATE PAID:	SUB TOTAL:
	TAX:
	INVOICE TOTAL:

Notes:_____

CLIENT NAME:				
ADDRESS:				
HOME PHONE: CELL PHONE:		EMAIL: OTHER:		

DATE	MATERIALS	TIME	$ HOUR	TOTAL $

COMPLETION DATE:	SUB TOTAL:
INVOICE NUMBER:	TAX:
DATE SENT:	
DATE PAID:	INVOICE TOTAL:

Notes:

CLIENT NAME:				
ADDRESS:				
HOME PHONE: CELL PHONE:		EMAIL: OTHER:		

DATE	MATERIALS	TIME	$ HOUR	TOTAL $

COMPLETION DATE:	SUB TOTAL:
INVOICE NUMBER:	TAX:
DATE SENT:	
DATE PAID:	INVOICE TOTAL:

Notes:

CLIENT NAME:

ADDRESS:

HOME PHONE: **EMAIL:**
CELL PHONE: **OTHER:**

DATE	MATERIALS	TIME	$ HOUR	TOTAL $

COMPLETION DATE: **SUB TOTAL:**

INVOICE NUMBER: **TAX:**

DATE SENT:

DATE PAID: **INVOICE TOTAL:**

Notes:_____
_____
_____
_____
_____
_____
_____
_____
_____
_____
_____
_____
_____
_____
_____
_____
_____
_____
_____
_____
_____
_____

CLIENT NAME:

ADDRESS:

HOME PHONE: EMAIL:
  CELL PHONE: OTHER:

DATE	MATERIALS	TIME	$ HOUR	TOTAL $

COMPLETION DATE:	SUB TOTAL:
INVOICE NUMBER:	TAX:
DATE SENT:	
DATE PAID:	INVOICE TOTAL:

Notes:_____

_____

_____

_____

_____

_____

_____

_____

_____

_____

_____

_____

_____

_____

_____

_____

_____

_____

_____

_____

_____

_____

CLIENT NAME:				
ADDRESS:				
HOME PHONE: CELL PHONE:		EMAIL: OTHER:		

DATE	MATERIALS	TIME	$ HOUR	TOTAL $

COMPLETION DATE:	SUB TOTAL:
INVOICE NUMBER:	TAX:
DATE SENT:	
DATE PAID:	INVOICE TOTAL:

Notes:

CLIENT NAME:				
ADDRESS:				
HOME PHONE: CELL PHONE:		EMAIL: OTHER:		

DATE	MATERIALS	TIME	$ HOUR	TOTAL $

COMPLETION DATE:	SUB TOTAL:
INVOICE NUMBER:	TAX:
DATE SENT:	
DATE PAID:	INVOICE TOTAL:

Notes:_____

CLIENT NAME:				
ADDRESS:				
HOME PHONE: CELL PHONE:		EMAIL: OTHER:		

DATE	MATERIALS	TIME	$ HOUR	TOTAL $

COMPLETION DATE:	SUB TOTAL:
INVOICE NUMBER:	TAX:
DATE SENT:	
DATE PAID:	INVOICE TOTAL:

Notes:_____
_____
_____
_____
_____
_____
_____
_____
_____
_____
_____
_____
_____
_____
_____
_____
_____
_____
_____
_____
_____
_____

CLIENT NAME:				
ADDRESS:				
HOME PHONE: CELL PHONE:		EMAIL: OTHER:		

DATE	MATERIALS	TIME	$ HOUR	TOTAL $

COMPLETION DATE:	SUB TOTAL:
INVOICE NUMBER:	TAX:
DATE SENT:	
DATE PAID:	INVOICE TOTAL:

Notes:

CLIENT NAME:				

ADDRESS:

HOME PHONE: **EMAIL:**
  CELL PHONE: **OTHER:**

DATE	MATERIALS	TIME	$ HOUR	TOTAL $

COMPLETION DATE: **SUB TOTAL:**

INVOICE NUMBER: **TAX:**

DATE SENT:

DATE PAID: **INVOICE TOTAL:**

Notes:

CLIENT NAME:

ADDRESS:

HOME PHONE: EMAIL:
  CELL PHONE: OTHER:

DATE	MATERIALS	TIME	$ HOUR	TOTAL $

COMPLETION DATE:	SUB TOTAL:
INVOICE NUMBER:	TAX:
DATE SENT:	
DATE PAID:	INVOICE TOTAL:

Notes:

CLIENT NAME:				
ADDRESS:				
HOME PHONE: CELL PHONE:		EMAIL: OTHER:		

DATE	MATERIALS	TIME	$ HOUR	TOTAL $

COMPLETION DATE:	SUB TOTAL:
INVOICE NUMBER:	TAX:
DATE SENT:	
DATE PAID:	INVOICE TOTAL:

Notes:

CLIENT NAME:				
ADDRESS:				
HOME PHONE: CELL PHONE:		EMAIL: OTHER:		

DATE	MATERIALS	TIME	$ HOUR	TOTAL $

COMPLETION DATE:	SUB TOTAL:
INVOICE NUMBER:	TAX:
DATE SENT:	
DATE PAID:	INVOICE TOTAL:

Notes:_____

CLIENT NAME:				
ADDRESS:				
HOME PHONE: CELL PHONE:		EMAIL: OTHER:		

DATE	MATERIALS	TIME	$ HOUR	TOTAL $

COMPLETION DATE:	SUB TOTAL:
INVOICE NUMBER:	TAX:
DATE SENT:	
DATE PAID:	INVOICE TOTAL:

Notes:_____

CLIENT NAME:				

ADDRESS:

HOME PHONE: **EMAIL:**
CELL PHONE: **OTHER:**

DATE	MATERIALS	TIME	$ HOUR	TOTAL $

COMPLETION DATE: **SUB TOTAL:**

INVOICE NUMBER:

DATE SENT: **TAX:**

DATE PAID: **INVOICE TOTAL:**

Notes:

CLIENT NAME:

ADDRESS:

HOME PHONE: **EMAIL:**
  CELL PHONE: **OTHER:**

DATE	MATERIALS	TIME	$ HOUR	TOTAL $

COMPLETION DATE: **SUB TOTAL:**

INVOICE NUMBER: **TAX:**

DATE SENT:

DATE PAID: **INVOICE TOTAL:**

Notes:_____
_____
_____
_____
_____
_____
_____
_____
_____
_____
_____
_____
_____
_____
_____
_____
_____
_____
_____
_____
_____

CLIENT NAME:				

ADDRESS:				

HOME PHONE:	EMAIL:
CELL PHONE:	OTHER:

DATE	MATERIALS	TIME	$ HOUR	TOTAL $

COMPLETION DATE:	SUB TOTAL:
INVOICE NUMBER:	TAX:
DATE SENT:	
DATE PAID:	INVOICE TOTAL:

Notes:_____

CLIENT NAME:				
ADDRESS:				
HOME PHONE: CELL PHONE:		EMAIL: OTHER:		

DATE	MATERIALS	TIME	$ HOUR	TOTAL $

COMPLETION DATE:	SUB TOTAL:
INVOICE NUMBER:	TAX:
DATE SENT:	
DATE PAID:	INVOICE TOTAL:

Notes:_____
_____
_____
_____
_____
_____
_____
_____
_____
_____
_____
_____
_____
_____
_____
_____
_____
_____
_____
_____
_____

	CLIENT NAME:			
	ADDRESS:			
	HOME PHONE: CELL PHONE:		EMAIL: OTHER:	

DATE	MATERIALS	TIME	$ HOUR	TOTAL $

COMPLETION DATE:	SUB TOTAL:
INVOICE NUMBER:	TAX:
DATE SENT:	
DATE PAID:	INVOICE TOTAL:

Notes:

CLIENT NAME:				
ADDRESS:				
HOME PHONE: CELL PHONE:		EMAIL: OTHER:		

DATE	MATERIALS	TIME	$ HOUR	TOTAL $

COMPLETION DATE:	SUB TOTAL:
INVOICE NUMBER:	TAX:
DATE SENT:	
DATE PAID:	INVOICE TOTAL:

Notes:

CLIENT NAME:				
ADDRESS:				
HOME PHONE: CELL PHONE:		EMAIL: OTHER:		

DATE	MATERIALS	TIME	$ HOUR	TOTAL $

COMPLETION DATE:	SUB TOTAL:
INVOICE NUMBER:	TAX:
DATE SENT:	
DATE PAID:	INVOICE TOTAL:

Notes:

CLIENT NAME:				

ADDRESS:

HOME PHONE: CELL PHONE:		EMAIL: OTHER:		

DATE	MATERIALS	TIME	$ HOUR	TOTAL $

COMPLETION DATE:	SUB TOTAL:
INVOICE NUMBER:	TAX:
DATE SENT:	
DATE PAID:	INVOICE TOTAL:

Notes:

CLIENT NAME:				
ADDRESS:				
HOME PHONE: CELL PHONE:		EMAIL: OTHER:		

DATE	MATERIALS	TIME	$ HOUR	TOTAL $

COMPLETION DATE:	SUB TOTAL:
INVOICE NUMBER:	TAX:
DATE SENT:	
DATE PAID:	INVOICE TOTAL:

Notes:

CLIENT NAME:

ADDRESS:

HOME PHONE: EMAIL:
 CELL PHONE: OTHER:

DATE	MATERIALS	TIME	$ HOUR	TOTAL $

COMPLETION DATE: SUB TOTAL:

INVOICE NUMBER:
                            TAX:
DATE SENT:

DATE PAID: INVOICE TOTAL:

Notes:

CLIENT NAME:				
ADDRESS:				
HOME PHONE: CELL PHONE:		EMAIL: OTHER:		

DATE	MATERIALS	TIME	$ HOUR	TOTAL $

COMPLETION DATE:	SUB TOTAL:
INVOICE NUMBER: DATE SENT:	TAX:
DATE PAID:	INVOICE TOTAL:

Notes:

CLIENT NAME:				
ADDRESS:				
HOME PHONE: CELL PHONE:		EMAIL: OTHER:		

DATE	MATERIALS	TIME	$ HOUR	TOTAL $

COMPLETION DATE:	SUB TOTAL:
INVOICE NUMBER: DATE SENT:	TAX:
DATE PAID:	INVOICE TOTAL:

Notes:

CLIENT NAME:				
ADDRESS:				
HOME PHONE: CELL PHONE:		EMAIL: OTHER:		

DATE	MATERIALS	TIME	$ HOUR	TOTAL $

COMPLETION DATE:	SUB TOTAL:
INVOICE NUMBER:	TAX:
DATE SENT:	
DATE PAID:	INVOICE TOTAL:

Notes:

CLIENT NAME:				
ADDRESS:				
HOME PHONE: CELL PHONE:		EMAIL: OTHER:		

DATE	MATERIALS	TIME	$ HOUR	TOTAL $

COMPLETION DATE:	SUB TOTAL:
INVOICE NUMBER:	TAX:
DATE SENT:	
DATE PAID:	INVOICE TOTAL:

Notes:

CLIENT NAME:				
ADDRESS:				
HOME PHONE: **CELL PHONE:**		EMAIL: OTHER:		

DATE	MATERIALS	TIME	$ HOUR	TOTAL $

COMPLETION DATE:	SUB TOTAL:
INVOICE NUMBER:	TAX:
DATE SENT:	
DATE PAID:	INVOICE TOTAL:

Notes:

CLIENT NAME:				
ADDRESS:				
HOME PHONE: **CELL PHONE:**		**EMAIL:** **OTHER:**		

DATE	MATERIALS	TIME	$ HOUR	TOTAL $

COMPLETION DATE:	SUB TOTAL:
INVOICE NUMBER:	TAX:
DATE SENT:	
DATE PAID:	INVOICE TOTAL:

Notes:

CLIENT NAME:				
ADDRESS:				
HOME PHONE: CELL PHONE:		EMAIL: OTHER:		

DATE	MATERIALS	TIME	$ HOUR	TOTAL $

COMPLETION DATE:	SUB TOTAL:
INVOICE NUMBER:	TAX:
DATE SENT:	
DATE PAID:	INVOICE TOTAL:

Notes:

CLIENT NAME:				

ADDRESS:				

HOME PHONE: CELL PHONE:		EMAIL: OTHER:		

DATE	MATERIALS	TIME	$ HOUR	TOTAL $

COMPLETION DATE:	SUB TOTAL:
INVOICE NUMBER:	TAX:
DATE SENT:	
DATE PAID:	INVOICE TOTAL:

Notes:

CLIENT NAME:				
ADDRESS:				
HOME PHONE: CELL PHONE:		EMAIL: OTHER:		

DATE	MATERIALS	TIME	$ HOUR	TOTAL $

COMPLETION DATE:	SUB TOTAL:
INVOICE NUMBER:	TAX:
DATE SENT:	
DATE PAID:	INVOICE TOTAL:

Notes:

CLIENT NAME:				
ADDRESS:				
HOME PHONE: CELL PHONE:		EMAIL: OTHER:		

DATE	MATERIALS	TIME	$ HOUR	TOTAL $

COMPLETION DATE:	SUB TOTAL:
INVOICE NUMBER:	TAX:
DATE SENT:	
DATE PAID:	INVOICE TOTAL:

Notes:

CLIENT NAME:

ADDRESS:

HOME PHONE: **EMAIL:**
  CELL PHONE: **OTHER:**

DATE	MATERIALS	TIME	$ HOUR	TOTAL $

COMPLETION DATE:	**SUB TOTAL:**
INVOICE NUMBER:	**TAX:**
DATE SENT:	
DATE PAID:	**INVOICE TOTAL:**

Notes:

	CLIENT NAME:			
	ADDRESS:			
	HOME PHONE: CELL PHONE:		EMAIL: OTHER:	

DATE	MATERIALS	TIME	$ HOUR	TOTAL $

COMPLETION DATE:	SUB TOTAL:
INVOICE NUMBER:	TAX:
DATE SENT:	
DATE PAID:	INVOICE TOTAL:

Notes:

CLIENT NAME:

ADDRESS:

HOME PHONE: **EMAIL:**
  CELL PHONE: **OTHER:**

DATE	MATERIALS	TIME	$ HOUR	TOTAL $

COMPLETION DATE: **SUB TOTAL:**

INVOICE NUMBER: **TAX:**

DATE SENT:

DATE PAID: **INVOICE TOTAL:**

Notes:

CLIENT NAME:				

ADDRESS:

HOME PHONE: **EMAIL:**
  CELL PHONE: **OTHER:**

DATE	MATERIALS	TIME	$ HOUR	TOTAL $

COMPLETION DATE:	SUB TOTAL:
INVOICE NUMBER:	TAX:
DATE SENT:	
DATE PAID:	INVOICE TOTAL:

Notes:

CLIENT NAME:				
ADDRESS:				
HOME PHONE: CELL PHONE:		EMAIL: OTHER:		

DATE	MATERIALS	TIME	$ HOUR	TOTAL $

COMPLETION DATE:	SUB TOTAL:
INVOICE NUMBER:	TAX:
DATE SENT:	
DATE PAID:	INVOICE TOTAL:

Notes:_____

CLIENT NAME:				

ADDRESS:

HOME PHONE: **EMAIL:**
CELL PHONE: **OTHER:**

DATE	MATERIALS	TIME	$ HOUR	TOTAL $

COMPLETION DATE:	SUB TOTAL:
INVOICE NUMBER:	TAX:
DATE SENT:	
DATE PAID:	INVOICE TOTAL:

Notes:_____

_____

_____

_____

_____

_____

_____

_____

_____

_____

_____

_____

_____

_____

_____

_____

_____

_____

_____

_____

CLIENT NAME:				
ADDRESS:				
HOME PHONE: CELL PHONE:		EMAIL: OTHER:		

DATE	MATERIALS	TIME	$ HOUR	TOTAL $

COMPLETION DATE:	SUB TOTAL:
INVOICE NUMBER:	TAX:
DATE SENT:	
DATE PAID:	INVOICE TOTAL:

Notes:

CLIENT NAME:

ADDRESS:

HOME PHONE: EMAIL:
  CELL PHONE: OTHER:

DATE	MATERIALS	TIME	$ HOUR	TOTAL $

COMPLETION DATE:	SUB TOTAL:
INVOICE NUMBER:	TAX:
DATE SENT:	
DATE PAID:	INVOICE TOTAL:

Notes:

CLIENT NAME:

ADDRESS:

HOME PHONE: EMAIL:
CELL PHONE: OTHER:

DATE	MATERIALS	TIME	$ HOUR	TOTAL $

COMPLETION DATE: SUB TOTAL:

INVOICE NUMBER: TAX:
DATE SENT:

DATE PAID: INVOICE TOTAL:

Notes:

CLIENT NAME:				
ADDRESS:				
HOME PHONE: CELL PHONE:		EMAIL: OTHER:		

DATE	MATERIALS	TIME	$ HOUR	TOTAL $

COMPLETION DATE:	SUB TOTAL:
INVOICE NUMBER:	TAX:
DATE SENT:	
DATE PAID:	INVOICE TOTAL:

Notes:_____

_____

_____

_____

_____

_____

_____

_____

_____

_____

_____

_____

_____

_____

_____

_____

_____

_____

_____

_____

_____

CLIENT NAME:				
ADDRESS:				
HOME PHONE: CELL PHONE:		EMAIL: OTHER:		

DATE	MATERIALS	TIME	$ HOUR	TOTAL $

COMPLETION DATE:	SUB TOTAL:
INVOICE NUMBER:	TAX:
DATE SENT:	
DATE PAID:	INVOICE TOTAL:

Notes:

CLIENT NAME:

ADDRESS:

HOME PHONE: EMAIL:
CELL PHONE: OTHER:

DATE	MATERIALS	TIME	$ HOUR	TOTAL $

COMPLETION DATE: SUB TOTAL:

INVOICE NUMBER:
                                  TAX:
DATE SENT:

DATE PAID: INVOICE TOTAL:

Notes:

CLIENT NAME:				
ADDRESS:				
HOME PHONE: CELL PHONE:		EMAIL: OTHER:		

DATE	MATERIALS	TIME	$ HOUR	TOTAL $

COMPLETION DATE:	SUB TOTAL:
INVOICE NUMBER: DATE SENT:	TAX:
DATE PAID:	INVOICE TOTAL:

Notes:_____
_____
_____
_____
_____
_____
_____
_____
_____
_____
_____
_____
_____
_____
_____
_____
_____
_____
_____
_____
_____
_____
_____

CLIENT NAME:				
ADDRESS:				
HOME PHONE: CELL PHONE:		EMAIL: OTHER:		

DATE	MATERIALS	TIME	$ HOUR	TOTAL $

COMPLETION DATE:	SUB TOTAL:
INVOICE NUMBER:	TAX:
DATE SENT:	
DATE PAID:	INVOICE TOTAL:

Notes:

CLIENT NAME:

ADDRESS:

HOME PHONE: EMAIL:
CELL PHONE: OTHER:

DATE	MATERIALS	TIME	$ HOUR	TOTAL $

COMPLETION DATE: SUB TOTAL:

INVOICE NUMBER:

DATE SENT: TAX:

DATE PAID: INVOICE TOTAL:

Notes:_____

_____

_____

_____

_____

_____

_____

_____

_____

_____

_____

_____

_____

_____

_____

_____

_____

_____

_____

_____

_____

CLIENT NAME:

ADDRESS:

HOME PHONE: **EMAIL:**
 CELL PHONE: **OTHER:**

DATE	MATERIALS	TIME	$ HOUR	TOTAL $

COMPLETION DATE: **SUB TOTAL:**

INVOICE NUMBER:

DATE SENT: **TAX:**

DATE PAID: **INVOICE TOTAL:**

Notes:

CLIENT NAME:				
ADDRESS:				
HOME PHONE: CELL PHONE:		EMAIL: OTHER:		

DATE	MATERIALS	TIME	$ HOUR	TOTAL $

COMPLETION DATE:	SUB TOTAL:
INVOICE NUMBER: DATE SENT:	TAX:
DATE PAID:	INVOICE TOTAL:

Notes:

CLIENT NAME:				

ADDRESS:

HOME PHONE: EMAIL:
CELL PHONE: OTHER:

DATE	MATERIALS	TIME	$ HOUR	TOTAL $

COMPLETION DATE: SUB TOTAL:

INVOICE NUMBER:
DATE SENT: TAX:

DATE PAID: INVOICE TOTAL:

Notes:

CLIENT NAME:

ADDRESS:

HOME PHONE: **EMAIL:**
  CELL PHONE: **OTHER:**

DATE	MATERIALS	TIME	$ HOUR	TOTAL $

COMPLETION DATE: **SUB TOTAL:**

INVOICE NUMBER:

DATE SENT: **TAX:**

DATE PAID: **INVOICE TOTAL:**

Notes:

CLIENT NAME:				
ADDRESS:				
HOME PHONE: CELL PHONE:		EMAIL: OTHER:		

DATE	MATERIALS	TIME	$ HOUR	TOTAL $

COMPLETION DATE:	SUB TOTAL:
INVOICE NUMBER: DATE SENT:	TAX:
DATE PAID:	INVOICE TOTAL:

Notes:

CLIENT NAME:				

ADDRESS:				

HOME PHONE: CELL PHONE:		EMAIL: OTHER:		

DATE	MATERIALS	TIME	$ HOUR	TOTAL $

COMPLETION DATE:	SUB TOTAL:
INVOICE NUMBER: DATE SENT:	TAX:
DATE PAID:	INVOICE TOTAL:

Notes:_____
_____
_____
_____
_____
_____
_____
_____
_____
_____
_____
_____
_____
_____
_____
_____
_____
_____
_____
_____
_____

CLIENT NAME:

ADDRESS:

HOME PHONE: EMAIL:
 CELL PHONE: OTHER:

DATE	MATERIALS	TIME	$ HOUR	TOTAL $

COMPLETION DATE: SUB TOTAL:

INVOICE NUMBER:
                                TAX:
DATE SENT:

DATE PAID: INVOICE TOTAL:

Notes:

CLIENT NAME:				
ADDRESS:				
HOME PHONE: CELL PHONE:		EMAIL: OTHER:		

DATE	MATERIALS	TIME	$ HOUR	TOTAL $

COMPLETION DATE:	SUB TOTAL:
INVOICE NUMBER:	TAX:
DATE SENT:	
DATE PAID:	INVOICE TOTAL:

Notes:_____

CLIENT NAME:				
ADDRESS:				
HOME PHONE: CELL PHONE:		EMAIL: OTHER:		

DATE	MATERIALS	TIME	$ HOUR	TOTAL $

COMPLETION DATE:	SUB TOTAL:	
INVOICE NUMBER:	TAX:	
DATE SENT:		
DATE PAID:	INVOICE TOTAL:	

Notes:

CLIENT NAME:				
ADDRESS:				
HOME PHONE: CELL PHONE:		EMAIL: OTHER:		

DATE	MATERIALS	TIME	$ HOUR	TOTAL $

COMPLETION DATE:	SUB TOTAL:
INVOICE NUMBER: DATE SENT:	TAX:
DATE PAID:	INVOICE TOTAL:

Notes:

CLIENT NAME:				
ADDRESS:				
HOME PHONE: CELL PHONE:		EMAIL: OTHER:		

DATE	MATERIALS	TIME	$ HOUR	TOTAL $

COMPLETION DATE:	SUB TOTAL:
INVOICE NUMBER:	TAX:
DATE SENT:	
DATE PAID:	INVOICE TOTAL:

Notes:

CLIENT NAME:

ADDRESS:

HOME PHONE: EMAIL:
  CELL PHONE: OTHER:

DATE	MATERIALS	TIME	$ HOUR	TOTAL $

COMPLETION DATE: SUB TOTAL:

INVOICE NUMBER: TAX:

DATE SENT:

DATE PAID: INVOICE TOTAL:

Notes:

CLIENT NAME:				
ADDRESS:				
HOME PHONE: CELL PHONE:		EMAIL: OTHER:		

DATE	MATERIALS	TIME	$ HOUR	TOTAL $

COMPLETION DATE:	SUB TOTAL:
INVOICE NUMBER:	TAX:
DATE SENT:	
DATE PAID:	INVOICE TOTAL:

Notes:

CLIENT NAME:				
ADDRESS:				
HOME PHONE: CELL PHONE:		EMAIL: OTHER:		

DATE	MATERIALS	TIME	$ HOUR	TOTAL $

COMPLETION DATE:	SUB TOTAL:
INVOICE NUMBER:	TAX:
DATE SENT:	
DATE PAID:	INVOICE TOTAL:

Notes:

CLIENT NAME:				
ADDRESS:				
HOME PHONE: CELL PHONE:		EMAIL: OTHER:		

DATE	MATERIALS	TIME	$ HOUR	TOTAL $

COMPLETION DATE:	SUB TOTAL:
INVOICE NUMBER:	TAX:
DATE SENT:	
DATE PAID:	INVOICE TOTAL:

Notes:

CLIENT NAME:				
ADDRESS:				
HOME PHONE: CELL PHONE:		EMAIL: OTHER:		

DATE	MATERIALS	TIME	$ HOUR	TOTAL $

COMPLETION DATE:	SUB TOTAL:
INVOICE NUMBER:	TAX:
DATE SENT:	
DATE PAID:	INVOICE TOTAL:

Notes:_____

CLIENT NAME:				
ADDRESS:				
HOME PHONE: CELL PHONE:		EMAIL: OTHER:		

DATE	MATERIALS	TIME	$ HOUR	TOTAL $

COMPLETION DATE:	SUB TOTAL:
INVOICE NUMBER:	TAX:
DATE SENT:	
DATE PAID:	INVOICE TOTAL:

Notes:

CLIENT NAME:

ADDRESS:

HOME PHONE: EMAIL:
CELL PHONE: OTHER:

DATE	MATERIALS	TIME	$ HOUR	TOTAL $

COMPLETION DATE: SUB TOTAL:

INVOICE NUMBER: TAX:

DATE SENT:

DATE PAID: INVOICE TOTAL:

Notes:

CLIENT NAME:				

ADDRESS:

HOME PHONE: **EMAIL:**
CELL PHONE: **OTHER:**

DATE	MATERIALS	TIME	$ HOUR	TOTAL $

COMPLETION DATE: **SUB TOTAL:**
INVOICE NUMBER: **TAX:**
DATE SENT:
DATE PAID: **INVOICE TOTAL:**

Notes:

CLIENT NAME:				
ADDRESS:				
HOME PHONE: CELL PHONE:		EMAIL: OTHER:		

DATE	MATERIALS	TIME	$ HOUR	TOTAL $

COMPLETION DATE:	SUB TOTAL:
INVOICE NUMBER: DATE SENT:	TAX:
DATE PAID:	INVOICE TOTAL:

Notes:

CLIENT NAME:

ADDRESS:

HOME PHONE: EMAIL:
CELL PHONE: OTHER:

DATE	MATERIALS	TIME	$ HOUR	TOTAL $

COMPLETION DATE: SUB TOTAL:

INVOICE NUMBER:
                                     TAX:
DATE SENT:

DATE PAID: INVOICE TOTAL:

Notes:_____
_____
_____
_____
_____
_____
_____
_____
_____
_____
_____
_____
_____
_____
_____
_____
_____
_____
_____

CLIENT NAME:				
ADDRESS:				
HOME PHONE: CELL PHONE:		EMAIL: OTHER:		

DATE	MATERIALS	TIME	$ HOUR	TOTAL $

COMPLETION DATE:	SUB TOTAL:
INVOICE NUMBER: DATE SENT:	TAX:
DATE PAID:	INVOICE TOTAL:

Notes:_____

_____

_____

_____

_____

_____

_____

_____

_____

_____

_____

_____

_____

_____

_____

_____

_____

_____

_____

_____

_____

CLIENT NAME:				
ADDRESS:				
HOME PHONE: CELL PHONE:		EMAIL: OTHER:		

DATE	MATERIALS	TIME	$ HOUR	TOTAL $

COMPLETION DATE:	SUB TOTAL:
INVOICE NUMBER:	TAX:
DATE SENT:	
DATE PAID:	INVOICE TOTAL:

Notes:

CLIENT NAME:				
ADDRESS:				
HOME PHONE: CELL PHONE:		EMAIL: OTHER:		

DATE	MATERIALS	TIME	$ HOUR	TOTAL $

COMPLETION DATE:	SUB TOTAL:
INVOICE NUMBER:	TAX:
DATE SENT:	
DATE PAID:	INVOICE TOTAL:

Notes:

CLIENT NAME:				
ADDRESS:				
HOME PHONE: CELL PHONE:		EMAIL: OTHER:		

DATE	MATERIALS	TIME	$ HOUR	TOTAL $

COMPLETION DATE:	SUB TOTAL:
INVOICE NUMBER:	TAX:
DATE SENT:	
DATE PAID:	INVOICE TOTAL:

Notes:

CLIENT NAME:				
ADDRESS:				
HOME PHONE: CELL PHONE:		EMAIL: OTHER:		

DATE	MATERIALS	TIME	$ HOUR	TOTAL $

COMPLETION DATE:	SUB TOTAL:
INVOICE NUMBER: DATE SENT:	TAX:
DATE PAID:	INVOICE TOTAL:

Notes:

CLIENT NAME:				
ADDRESS:				
HOME PHONE: CELL PHONE:		EMAIL: OTHER:		

DATE	MATERIALS	TIME	$ HOUR	TOTAL $

COMPLETION DATE:	SUB TOTAL:
INVOICE NUMBER: DATE SENT:	TAX:
DATE PAID:	INVOICE TOTAL:

Notes:_____
_____
_____
_____
_____
_____
_____
_____
_____
_____
_____
_____
_____
_____
_____
_____
_____
_____
_____
_____
_____

CLIENT NAME:				

ADDRESS:

HOME PHONE: EMAIL:
 CELL PHONE: OTHER:

DATE	MATERIALS	TIME	$ HOUR	TOTAL $

COMPLETION DATE: | SUB TOTAL:
INVOICE NUMBER: | TAX:
DATE SENT: |
DATE PAID: | INVOICE TOTAL:

Notes:_____
_____
_____
_____
_____
_____
_____
_____
_____
_____
_____
_____
_____
_____
_____
_____
_____
_____
_____
_____
_____
_____
_____

	CLIENT NAME:			
	ADDRESS:			
	HOME PHONE: CELL PHONE:	EMAIL: OTHER:		

DATE	MATERIALS	TIME	$ HOUR	TOTAL $

COMPLETION DATE:	SUB TOTAL:
INVOICE NUMBER:	TAX:
DATE SENT:	
DATE PAID:	INVOICE TOTAL:

Notes:

CLIENT NAME:				
ADDRESS:				
HOME PHONE: CELL PHONE:		EMAIL: OTHER:		

DATE	MATERIALS	TIME	$ HOUR	TOTAL $

COMPLETION DATE:	SUB TOTAL:
INVOICE NUMBER: DATE SENT:	TAX:
DATE PAID:	INVOICE TOTAL:

Notes:_____

_____

_____

_____

_____

_____

_____

_____

_____

_____

_____

_____

_____

_____

_____

_____

_____

_____

_____

_____

_____

CLIENT NAME:				

ADDRESS:

HOME PHONE: **EMAIL:**
  CELL PHONE: **OTHER:**

DATE	MATERIALS	TIME	$ HOUR	TOTAL $

COMPLETION DATE: **SUB TOTAL:**

INVOICE NUMBER: **TAX:**

DATE SENT:

DATE PAID: **INVOICE TOTAL:**

Notes:

CLIENT NAME:				
ADDRESS:				
HOME PHONE: CELL PHONE:		EMAIL: OTHER:		

DATE	MATERIALS	TIME	$ HOUR	TOTAL $

COMPLETION DATE:	SUB TOTAL:
INVOICE NUMBER:	TAX:
DATE SENT:	
DATE PAID:	INVOICE TOTAL:

Notes:

CLIENT NAME:				

ADDRESS:				

HOME PHONE: CELL PHONE:		EMAIL: OTHER:		

DATE	MATERIALS	TIME	$ HOUR	TOTAL $

COMPLETION DATE:	SUB TOTAL:
INVOICE NUMBER:	TAX:
DATE SENT:	
DATE PAID:	INVOICE TOTAL:

Notes:

Made in United States
Orlando, FL
25 March 2022

16131630R00114